**本课题受到以下项目资助**

◎ 中国科学院国际合作局项目
 "支持GRC开放获取行动计划的世界开放获取行动评价与监测平台"（Y140321001）
◎ 中国科学院科学传播局、发展规划局项目
 "中国科学院文献情报和期刊出版领域引进优秀人才计划"择优支持项目（Y14341001）
◎ 国家自然科学基金委项目
 "开放获取重大趋势与科学基金相关政策研究"（ZK16041001）

# 预印本平台
# （arXiv.org）简介

 >>> Oya Yildirim Rieger **主审**
顾立平 **编**

· 北京 ·

### 图书在版编目（CIP）数据

预印本平台（arXiv.org）简介 / 顾立平编. —北京：科学技术文献出版社，2016.9（2017.10 重印）
ISBN 978-7-5189-1880-5

Ⅰ.①预… Ⅱ.①顾… Ⅲ.①互联网络—应用—科学技术—信息资源—信息管理—介绍 Ⅳ.①TP393.092

中国版本图书馆 CIP 数据核字（2016）第 216251 号

## 预印本平台（arXiv.org）简介

策划编辑：崔灵菲　责任编辑：王瑞瑞　责任校对：张吲哚　责任出版：张志平

| | | |
|---|---|---|
| 出　版　者 | 科学技术文献出版社 | |
| 地　　　址 | 北京市复兴路15号　邮编 100038 | |
| 编　务　部 | （010）58882938，58882087（传真） | |
| 发　行　部 | （010）58882868，58882874（传真） | |
| 邮　购　部 | （010）58882873 | |
| 官　方　网　址 | www.stdp.com.cn | |
| 发　行　者 | 科学技术文献出版社发行　全国各地新华书店经销 | |
| 印　刷　者 | 虎彩印艺股份有限公司 | |
| 版　　　次 | 2016 年 9 月第 1 版　2017 年 10 月第 3 次印刷 | |
| 开　　　本 | 850×1168　1/32 | |
| 字　　　数 | 57 千 | |
| 印　　　张 | 3.5 | |
| 书　　　号 | ISBN 978-7-5189-1880-5 | |
| 定　　　价 | 42.00 元 | |

版权所有　违法必究

购买本社图书，凡字迹不清、缺页、倒页、脱页者，本社发行部负责调换

# 前 言

预印本平台（arXiv.org）为全世界的作者和研究人员提供了一个科学研究的开放获取知识库，对所有用户免费开放。始于1991年8月，由Paul Ginsparg创建，2001年将平台的运营、编辑、经济和管理转移到康奈尔大学图书馆。如今，它不仅改变了物理学多个领域的学术交流方式，而且在数学、计算机科学、定量生物学、定量金融学和统计学等领域发挥着越来越突出的作用。本书介绍arXiv.org的经营管理模式、用户需求调查及发展方向。

本书详细具体地介绍了arXiv的创建与发展、参与arXiv的准则，以及为减少arXiv的经济负担所提出的可持续发展倡议等。详细介绍了arXiv的收入来源与运营成本，以及成为会员的好处和方法流程。完整说明了arXiv的管理模式，重点介绍了其在康奈尔大学图书馆的运营，包括人员配备、保存策略、学科扩展等；详细解答了关于捐款和资助的各种问题。最后结合时间轴回顾了2015年arXiv愿景设置和资金筹集的策略，并结合用户调查报告，总结用户对arXiv的意见和建议，

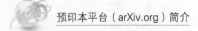 预印本平台（arXiv.org）简介

分析调整和改进的方向。本书中英文结合，相对全面地介绍了 arXiv 的组织、功能、已有成果、发展方向等。

　　本书知识内容的贡献者是 Oya Yildirim Rieger 女士。顾立平负责统筹编译、校对；最终稿件由 Oya Yildirim Rieger 主审。赵越协助本书绝大多数内容的编译，段美珍和王楠协助若干内容的编译和校对，杨良斌通读本书全文，史盈盈和丁利芳参与编辑。本书若有任何缺失、不足之处，应属编者顾立平之责任。

# Content

1  Genral Questions .................................................................. 1
   1.1  What Is arXiv ............................................................... 1
   1.2  arXiv's Key Principles ..................................................... 2
   1.3  How Are Submissions to arXiv Moderated ..................... 2

2  Sustainability Initiative ........................................................ 3
   2.1  Purpose ......................................................................... 3
   2.2  Outcomes ...................................................................... 3

3  arXiv Membership Model .................................................... 5
   3.1  The Goals of the New arXiv Membership Program ...... 5
   3.2  The Revenue Model for arXiv ........................................ 5
   3.3  arXiv's Operating Costs ................................................. 6
   3.4  The Institutional Membership Fees ................................ 7
   3.5  The Benefits of Becoming a Member ............................ 8
   3.6  Focus on the Top 200 Institutional Users ...................... 9
   3.7  How Are the Member Institutions Acknowledged ......... 9
   3.8  Is the Tier Model Fair .................................................... 10

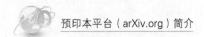
预印本平台（arXiv.org）简介

  3.9 Membership Pledges ............................................................ 10

4 arXiv Governance Model ........................................................... 13
  4.1 How Does the arXiv Governance Model Work .......... 13
  4.2 When Will the New Model Begin ................................ 15

5 Staffing and Expenses ............................................................... 16
  5.1 arXiv Staffing Projections and the Projected Expenses for 2013-2017 ................................................................ 16
  5.2 Additional Revenue Sources ........................................ 16
  5.3 The Relationship Between arXiv and SCOAP3 .......... 17

6 arXiv @ CUL ............................................................................... 19
  6.1 How Is arXiv Staffed ...................................................... 19
  6.2 Does arXiv Accept Research Data ............................... 19
  6.3 CUL's Preservation Strategies ..................................... 20
  6.4 Subjects Expansion ....................................................... 21
  6.5 How to Stay Informed about arXiv and Download Papers Automatically ................................................... 22

7 Personal Donations .................................................................... 23
  7.1 Donation Methods ........................................................ 23
  7.2 Initiatives Supported .................................................... 23
  7.3 Other Questions about Donation ................................ 24

# Content

**8 arXiv Review Strategy** ............................................. 27
    8.1   Abstract ................................................................. 27
    8.2   arXiv@25: Strategy and Timeline for Vision Setting and Fund Raising .............................................. 28

**9 arXiv User Survey Report** .......................................... 31
    9.1   Executive Summary ............................................. 31
    9.2   Key Findings ........................................................ 34
    Appendix A: Demographics of Respondents .............. 43
    Appendix B: Opinions on arXiv's Current Services & Future Directions ................................................. 47

# 目 录

1 arXiv 的一般性问题 ............................................. 56
   1.1   arXiv 是什么 ............................................. 56
   1.2   arXiv 的主要准则 ......................................... 57
   1.3   提交文章的合适方法 ..................................... 57

2 可持续发展倡议 .................................................. 58
   2.1   倡议的目的 ............................................... 58
   2.2   结果 ..................................................... 58

3 arXiv 的会员模式 ................................................ 60
   3.1   arXiv 会员项目的目标 .................................... 60
   3.2   arXiv 的收入模式 ........................................ 60
   3.3   arXiv 的运营成本 ........................................ 61
   3.4   机构会员的费用 .......................................... 62
   3.5   成为会员的好处 .......................................... 62
   3.6   重点关注前 200 名机构用户 ............................... 63
   3.7   成员机构的认可方式 ..................................... 63
   3.8   模式是否公平 ............................................ 63
   3.9   会员承诺 ................................................. 64

## 4 arXiv 的管理模式 .................................................. 66
### 4.1 管理模式的运作 ............................................. 66
### 4.2 新管理模式的启动时间 ....................................... 68

## 5 职员和费用 ...................................................... 69
### 5.1 arXiv 2013—2017 年的职员预测和支出预测 ..................... 69
### 5.2 附加收入来源 ............................................... 69
### 5.3 arXiv 和 SCOAP3 之间的关系 ................................. 69

## 6 arXiv 在康奈尔大学图书馆 ......................................... 71
### 6.1 人员配备 ................................................... 71
### 6.2 接收研究数据 ............................................... 71
### 6.3 保存策略 ................................................... 72
### 6.4 学科扩展 ................................................... 72
### 6.5 了解 arXiv 和自动下载论文的方法 ............................. 73

## 7 个人捐款 ........................................................ 74
### 7.1 捐款方式 ................................................... 74
### 7.2 支持项目 ................................................... 74
### 7.3 关于捐款的其他问题 ......................................... 74

## 8 arXiv 策略回顾 .................................................. 78
### 8.1 概要 ....................................................... 78
### 8.2 愿景设置和资金筹集的策略与时间轴 ........................... 78

# 目 录

9 arXiv 用户调查报告 ...................................................... 81
  9.1 执行摘要 ........................................................... 81
  9.2 主要的调查结果 ..................................................... 83
  附录 A：调查对象统计 .................................................... 90
  附录 B：对 arXiv 目前服务与未来方向的观点 ................ 94

# 1 Genral Questions

## 1.1 What Is arXiv

arXiv.org is widely acknowledged as one of the most successful open-access digital archives. In August 1991, Paul Ginsparg created arXiv as a repository for preprints in physics; in 2001, it moved to Cornell University Library (CUL). Now, it has transformed scholarly communication in multiple fields of physics and plays an increasingly prominent role in mathematics, computer science, quantitative biology, quantitative finance, and statistics. It is embedded in the research workflows of these subjects and allows the rapid dissemination of scientific findings. By providing open access to researchers worldwide, arXiv makes science more democratic.

As of August 2012, arXiv contains more than 770,000 e-prints. In 2011, the repository saw 76,578 new submissions and close to 50 million downloads from all over the world. arXiv is international in scope, with mirror sites in nine countries and collaborations

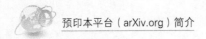

with U.S. and foreign professional societies and other international organizations.

## 1.2 arXiv's Key Principles

arXiv provides an open-access repository of scientific research to authors and researchers worldwide. It is a moderated scholarly communication forum informed and guided by scientists and the scientific cultures it serves. Access via arXiv.org is free to all end-users, and individual researchers can deposit their own content in arXiv for free. Please see the arXiv Principles document for the operational, editorial, economic, and governance principles.

## 1.3 How Are Submissions to arXiv Moderated

Material submitted to arXiv is expected to be interesting, relevant, and valuable to those disciplines. arXiv reserves the right to reject or reclassify any submission. Expert moderators review submissions to verify that they are topical and referable scientific contributions that follow accepted standards of scholarly communication (as exemplified by conventional journal articles). Please see http://arxiv.org/help for information about submission and review policies.

# 2 Sustainability Initiative

## 2.1 Purpose

Since 2010, Cornell University Library (CUL)'s sustainability planning initiative has aimed to reduce arXiv's financial burden and dependence on a single institution, instead creating a broad-based, community-supported resource. To keep an open-access academic resource like arXiv sustainable, administrators must continue to enhance its value based on the needs of all its user communities, as well as cover the associated operating costs. Cornell's initiative strived to strengthen arXiv's technical, service, financial, and policy infrastructure.

## 2.2 Outcomes

As an interim strategy, CUL initially established a voluntary institutional contribution model, inviting pledges from 200 libraries and research laboratories worldwide that represent arXiv's heaviest institutional users. During this 2010-2012 planning process, CUL sought input from a wide range of stakeholders to position arXiv as a collaboratively governed, community-supported resource. The planning effort also included a meeting with representatives of several publishers and societies to discuss the feasibility and desirability of collaborating, which would improve crosslinking, interoperability, and lifecycle support for research material.

The sustainability planning process aimed to investigate how to diversify revenue models, ensure that arXiv meets a set of governance principles, and provide transparent and reliable community-supported service. arXiv's membership model is the key outcome of the three-year planning process, which was facilitated by a planning grant from the Simons Foundation. The background information about the planning activities is available at http://arxiv.org/help/support.

# 3 arXiv Membership Model

## 3.1 The Goals of the New arXiv Membership Program

To help with arXiv's support and governance, the membership program aims to engage libraries and research laboratories worldwide that represent arXiv's heaviest institutional users. Each member institution pledges a five-year initial funding commitment to support arXiv. arXiv's sustainability plan is founded on arXiv's operating principles and presents a business model for generating revenues.

## 3.2 The Revenue Model for arXiv

Cornell University Library (CUL), the Simons Foundation, and a global collective of institutional members support arXiv financially. The financial model for 2013-2017 entails three sources of revenues:

- CUL provides a cash subsidy of $75,000 per year in support of arXiv's operational costs and an in-kind contribution of all indirect costs, which currently represents 37% of total operating expenses.

- The Simons Foundation contributes $50,000 per year in recognition of CUL's stewardship of arXiv. In addition, the Foundation matches $300,000 per year of the funds generated through arXiv membership fees.

- Each member institution pledges a five-year funding commitment to support arXiv. Based on institutional usage ranking, the annual fees are set in four tiers from $1,500-$3,000. Cornell's goal is to raise $300,000 per year through membership fees generated by approximately 126 institutions.

The gift aims to encourage long-term community support by lowering arXiv membership fees, making participation affordable to a broader range of institutions. This support helps ensure that the ultimate responsibility for sustaining arXiv remains with the research communities and institutors that benefit from the service most directly.

## 3.3 arXiv's Operating Costs

arXiv's operating costs for 2013-2017 are projected to average of $826,000 per year, including indirect expenses. The operating budget projections for 2012-2017 include the four key sources of revenues mentioned above: Cornell's annual funding of $75,000 per year, plus indirect expenses; the $50,000 per year gift from the Simons Foundation; annual fee income from the member institutions; and the $300,000 per year challenge grant from the Simons Foundation, based on the revenues generated through membership payments.

The arXiv contingency fund supports unexpected expenses to ensure a sound business model. Currently, arXiv has a contingency fund of approximately $125,000 accumulated during the last two years due to unexpected staff vacancies and other savings, such as the transition to virtual servers. Guidelines for using these funds will be discussed with the Member Advisory Board.

## 3.4 The Institutional Membership Fees

Membership fees are based on an institutional ranking calculated according to the number of articles downloaded. Usage statistics and institutional rankings are calculated annually. The usage statistics for the top 200 institutional users for 2009-2011 are available online at arXiv Support. To encourage participation in the arXiv membership program, the fees will decrease as the number of participating institutions increases.

| Usage Rank | Annual Membership Fees |
|---|---|
| 1~50 | $3,000 |
| 51~100 | $2,500 |
| 101~150 | $2,000 |
| 151+ | $1,500 |

## 3.5 The Benefits of Becoming a Member

Exclusive benefits for participating organizations include:

● Participation in arXiv's ongoing governance through the Member Advisory Board, which provides input for project prioritization, new service offerings, financial planning, use of discretionary funds, future technical developments, and policy decisions;

● Access to enhanced institutional use statistics;

● Public acknowledgement of members'role in financial support.

Several other benefits are under consideration, including automatic posting of arXiv submissions to the institutional repository at an author's host institution and creating a members' portal to provide timely information for the participating institutions. Such potential benefits need to be explored further to understand delivery and maintenance requirements.

## 3.6 Focus on the Top 200 Institutional Users

Use is measured by article downloads. CUL focuses on the top 200 institutions because they account for about 75% of institutionally identifiable downloads. This is not a mandatory fee-based funding model that would force anyone to support arXiv to secure access to the content; access to arXiv both for submitters and users will remain free. arXiv is also be happy to accept and acknowledge support from other libraries and research laboratories.

## 3.7 How Are the Member Institutions Acknowledged

Sustainability initiative information, including a list of arXiv supporters, is available under the Help option from the main menu (http://arxiv.org/help/support). Through IP identification, the users of these institutions see an acknowledgement banner to indicate support from their home institutions. CUL is planning to make this information more prominent for arXiv's users, specifically to communicate participants' contributions to the institutional scientists. It is also critical to raise awareness about the resource needs for running scientific repositories and the roles of member libraries in securing arXiv's future.

## 3.8 Is the Tier Model Fair

arXiv is a primary destination site for authors and readers in its core domains within physics and math. arXiv's sustainability should be considered a shared investment in a culturally embedded resource that provides unambiguous value to a global network of science researchers. Any system of voluntary contribution is susceptible to free-riders, but arXiv is extremely cost-effective, so even modest contributions from heavy-user institutions will support continued open access for all while providing good value-for-money when compared with subscription services. The arXiv Membership Model offers exclusive benefits for participating organizations in recognition of their contributions to arXiv.

## 3.9 Membership Pledges

### 3.9.1 Who Can Be a Member

Membership in arXiv and representation on the Member Advisory Board (MAB) is reserved for libraries, research institutions, laboratories, and foundations that contribute to the financial support of the service.

### 3.9.2 The Process to Nominate Candidate

This information will be added in September 2012 at http://arxiv.org/help/support.

### 3.9.3 An Agreement for Five-year Pledges of Support

Cornell University Library (CUL) envisions the five-year commitment as a good-faith intention to support arXiv from 2013 to 2017. It is not seen as a lump-sum payment or a legally binding contract.

### 3.9.4 The Timeline for Invoicing

Institutions will be invoiced by March 31, 2013. Timelines are flexible, and for past supporters, CUL will retain the current invoicing schedule. Please contact us to request alternative invoice date.

If the institution which is not listed in the Top 200 wants to become a supporting member, please email us at support@arxiv.org.

### 3.9.5 Ask for Institutional IP Addresses

Members will be openly acknowledged on the arXiv.org website. For institutions that provide IP address information, CUL can customize the message in the upper right of the arXiv banner to read, "We gratefully acknowledge support from_____."

### 3.9.6 Find Information

arXiv's institutional statistics can be found online for 2009, 2010 and 2011. If your institution is not listed, please email us at support@arxiv.org for usage statistics.

We often receive questions about the feasibility of providing submission-based statistics and comparing them with our current institutional downloads statistics. Currently, the author metadata for arXiv is not sufficiently consistent to support any systematic analysis. We have conducted a manual analysis of a single month of submissions, and the results from that sampling indicate that submission- and download-based data exhibit similar characteristics in means of institutional ranking. We continue to refine our findings and appreciate ideas about how this information can be used for business-planning purposes. In time, arXiv expects to undertake metadata remediation to improve the authorship data for existing submissions. This will enable better author linking, allow improvements in ownership claiming, and pave the way to support author identity linking using the proposed ORCID author identifiers.

# 4 arXiv Governance Model

## 4.1 How Does the arXiv Governance Model Work

Cornell University Library (CUL) holds the overall administrative and financial responsibility for arXiv's operation and development, with guidance from its Member Advisory Board (MAB) and its Scientific Advisory Board (SAB). CUL manages the moderation of submissions and user support, including the development and implementation of policies and procedures; operates arXiv's technical infrastructure; assumes responsibility for archiving to ensure long-term access; oversees arXiv mirror sites; and establishes and maintains collaborations with related initiatives to improve services for the scientific community through interoperability and tool-sharing.

The SAB is composed of scientists and researchers in disciplines covered by arXiv. It provides advice and guidance pertaining to the repository's intellectual oversight, with a particular focus on the policies and operation of arXiv's moderation system.

SAB reviews and revises the criteria and standards for deposit in arXiv; proposes new subject or discipline domains to be covered by arXiv; provides scientists' feedback on arXiv development projects proposed by the MAB; and makes recommendations regarding development projects (in particular, suggestions about improving the systems that support submission and moderation processing).

The MAB is made up of elected representatives from arXiv's membership and serves as a consultative body. The board represents participating institutions'interests and advises CUL on issues related to repository management and development, standards implementation, interoperability, development priorities, business planning, financial planning, and outreach and advocacy. Membership in arXiv and representation on the MAB is reserved for libraries, research institutions, laboratories, and foundations that contribute to the financial support of the service. Voting for representatives to the MAB is open to all eligible member institutions, which each hold one vote. The MAB bylaws specify the board's operations, including composition, member appointment methods, and term of office.

arXiv already has a Scientific Advisory Board, which was established several years ago by Paul Ginsparg. Cornell's arXiv team is in the process of gathering SAB's input on specifics of composition, terms, and selection of board members in order to create bylaws for the group. CUL formed an interim MAB in April 2012 to review and ratify bylaws for the MAB that specify its operations, including member appointment methods, term of office, etc. The interim group also aims to define measures of success for arXiv and the new governance model, and it will test potential governance scenarios to better understand the dynamics between the CUL team, MAB, and SAB. This will determine respective responsibilities and refine the arXiv governance principles.

## 4.2 When Will the New Model Begin

The goal is to transition from the current three-year sustainability planning phase to a long-term model with clearly identified roles for CUL, the MAB, and the SAB by December 2012.

# 5  Staffing and Expenses

## 5.1  arXiv Staffing Projections and the Projected Expenses for 2013-2017

Staffing projections are contained in the Business Model document.

arXiv's budget is structured as a grant-funded account at CUL in order to adhere to Cornell's financial policies and to provide audit tracking. Projected arXiv expenses are described in the Business Model document.

## 5.2  Additional Revenue Sources

In the process of investigating business models, CUL considered many options and ruled out article-processing charges and submission fees. Barrier-free submission and use is one of the founding principles of arXiv. arXiv's advisory boards will consider adding other sources of revenue, such as including a "give" button on the arXiv webpage to encourage scientists and other interested parties to donate.

## 5.3 The Relationship Between arXiv and SCOAP3

arXiv and SCOAP3 (Sponsoring Consortium for Open Access Publishing in Particle Physics) are complementary in both scope and function. arXiv is a scholarly communication forum that serves the needs of researchers for rapid open-access dissemination of pre-prints in a broad range of fields including physics, mathematics, computer science, quantitative biology, quantitative finance, and statistics. SCOAP3 tackles the issue of sustainable open access to peer-reviewed publications in high-energy physics (HEP), one of the many categories supported by arXiv (and typically representing less than 20% of submissions to arXiv). These publications appear up to one year later than the arXiv preprint, and are important for the community as the definitive versions of record used by universities and funding agencies for accountability purposes, while the scholarly discourse proceeds on arXiv (see http://arxiv.org/abs/0906.5418).

Whereas SCOAP3 collaborates with publishers to provide open access to formally published, refereed HEP content, arXiv is a moderated scholarly communication forum informed and guided by scientists and the scientific cultures it serves. Expert moderators review submissions to verify that they are topical and of interest to the scientific community, follow accepted standards of scholarly communication, and are classified in the appropriate subject categories. As such, arXiv classification is a valuable and crucial element in the identification of articles for which SCOAP3 financially supports publication. As the SCOAP3 initiative continues to evolve, arXiv and SCOAP3 are fully committed to further discussions about shared opportunities in the area of open-access scientific communication, leveraging the strong co-operative bonds of the two teams in other initiatives. At this point, arXiv financial planning does not assume any direct transfer of resources from the SCOAP3 project.

# 6   arXiv @ CUL

## 6.1   How Is arXiv Staffed

arXiv's direct operating expenses include approximately 6 full-time employees (FTE). Over half of these staff are involved in supporting users and administering the moderation system. Another 1.5 to 2 FTE provide programming and system support, and there is a 0.5 FTE management component. Additional management and business support, such as the contributions of the arXiv Program Director, is covered under indirect expenses. For detailed five-year staffing projections, see the arXiv Organizational Structure.

## 6.2   Does arXiv Accept Research Data

In 2011, arXiv launched a pilot data upload interface for data associated with arXiv articles. Submission of data is accomplished through small extensions to arXiv's submission interface. While the article is announced and stored on arXiv, data is automatically deposited in the Data Conservancy repository and linked from the article (see http://arxiv.org/help/data_conservancy for more information). This is a pilot project that will be re-evaluated in collaboration with the Data Conservancy by the end of 2012. Small amounts of data, program code, etc. may also be stored as ancillary files alongside arXiv articles (see http://arxiv.org/help/ancillary_files for more information).

## 6.3 CUL's Preservation Strategies

Digital preservation refers to a range of managed activities to support the long-term maintenance of bitstream. These activities ensure that digital objects are usable (intact and readable), retaining all quantities of authenticity, accuracy, and functionality deemed to be essential when articles (and other associated materials) were ingested. Formats accepted by arXiv have been selected based on their archival value (TeX/LaTeX, PDF, HTML) and the ability to process all source files is actively monitored. The underlying bits are protected by standard backup procedures at the Cornell campus. Off-site backup facilities in New York City provide geographic redundancy. The complete content is replicated at arXiv's mirror sites around the world, and additional managed tape backups are taken at Los Alamos National Laboratory. CUL has an archival repository to support preservation of critical content from institutional resources, including arXiv. We anticipate storing all arXiv documents, both in source and processed form, in this repository. There will be ongoing incremental ingest of new material. We expect that CUL will bear the preservation costs for arXiv, leveraging the archival infrastructure developed for the library system.

### 6.4 Subjects Expansion

We have adopted a measured approach to expansion, because significant organizational and administrative efforts are required to create and maintain new subject areas. Adding a new subject area involves exploring the user base and use characteristics pertaining to the subject area, establishing the necessary advisory committees, and recruiting moderators. Also, although arXiv.org is the central portal for scientific communication in some disciplines, it is neither feasible nor necessarily desirable to play that role in all disciplines.

Although we anticipate that arXiv will become increasingly broad in its subject area coverage, we believe this development must occur in a planned and strategic manner. One of the arXiv principles is that any expansion into other subjects or disciplines must include scholarly community support, satisfy arXiv's quality standards, and take into consideration its operational capacity and financial requirements.

## 6.5 How to Stay Informed about arXiv and Download Papers Automatically

Please join the arXiv support announcement email list by sending an email message to arxiv-support-updates-L-request@cornell.edu. Leave the subject line blank. The body of the message should be a single word: join.

Please see details of Institutional Repository (IR) Interoperability.

# 7 Personal Donations

## 7.1 Donation Methods

You may donate to arXiv through the Cornell University Alumni & Friends website. 100% of your contribution will fund improvements and new initiatives that will benefit arXiv's global scientific community.

## 7.2 Initiatives Supported

Please see the Special Projects section in the arXiv roadmap. As we described in our last annual update, the current business model is working well in covering arXiv's baseline maintenance costs with support from 190 member organizations, the Simons Foundation, and Cornell University Library. However, we need to raise additional resources in order to fund new initiatives that are beyond the routine operational work.

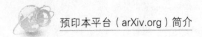
预印本平台（arXiv.org）简介

## 7.3 Other Questions about Donation

(1) What is arXiv's history and impact?

Please see this press release that highlights the significance of arXiv: arXiv Hits 1 Million Submissions (see https://www.library.cornell.edu/about/news/press-releases/arxiv-hits-1-million-submissions-0).

(2) What is arXiv's sustainability plan?

See our business model and sustainability plan for information about our current budget, governance model, priorities, and annual reports.

(3) What is arXiv's current operating budget?

See our budget.

(4) I thought I was supporting arXiv. Why does a payment page from Cornell University appear?

arXiv is managed by Cornell University Library. The workflow for processing our special projects is managed by Cornell Alumni Affairs and Development. Your gift will support arXiv and only arXiv.

(5) I am donating online, what is arXiv going to do with my personal information?

Your address and phone number will be used solely to verify

your credit card information, and will not be added to any contact lists. Neither arXiv.org nor Cornell retains this information.

(6) I am donating online, why am I being asked for "In Honor/Memory" information?

The online donation portal is managed by Cornell Alumni Affairs and Development and has several fields not required to donate to arXiv.org, "In Honor/Memory" designations are amongst these. Please feel free to donate "In Honor/Memory" if you like, or to skip this designation by pressing the "Continue" button (on the Lower Right).

(7) Is my institution currently supporting arXiv?

See our donors.

(8) My institution currently supports arXiv. Why should I?

Our member organizations give generously to support our current operating budget. The current 5-year business plan represents a baseline maintenance scenario. It was developed based on an analysis of the arXiv's baseline expenses during 2010-2012. It does not factor in any new functionality requirements or other unforeseen resource needs. Although a development reserve was established to fund such expenses, it is not sufficient to subsidize significant development efforts through surplus funds. Your gift will help fund innovation and enhancements so arXiv can remain a

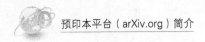

预印本平台（arXiv.org）简介

competitive, global open access resource.

(9) How can I stay up to date on arXiv?

Please send an email to arxiv-support-updates-L-request@cornell.edu:

- Leave the subject line blank.
- Body of the message: join.

(10) Is my gift tax deductible?

Yes.

(11) What is Cornell's tax ID number?

150532082.

(12) Do I get a receipt?

Yes. After you give you will be issued a thank you page. The thank you page serves as a receipt.

(13) Can I contribute from outside the U.S.?

Yes. Cornell accepts international credit cards, and arXiv welcomes support from around the world!

(14) I have other questions and want to talk with someone from the arXiv team.

You can contact us at support@cornell.edu. We welcome your questions and comments.

# 8 arXiv Review Strategy

## 8.1 Abstract

From users' perspective, arXiv continues to be a successful, prominent subject repository system serving the needs of many scientists around the world. However, under the hood, the service is facing significant pressures. The conclusion of the Scientific Advisory Board (SAB) and the Member Advisory Board (MAB) 2015 annual meetings was that, in addition to the current business model with a focus on maintenance, the arXiv team needs to embark on a significant fundraising effort, pursuing grants and collaborations. We need to first create a compelling and coherent vision to be able to persuasively articulate our fund raising goals beyond the current sustainability plan that aims to support the baseline operation. We'd like to use the approaching 25th anniversary of arXiv as an important milestone to engage us in a series of vision-setting exercises.

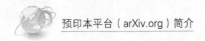
预印本平台（arXiv.org）简介

## 8.2 arXiv@25: Strategy and Timeline for Vision Setting and Fund Raising

(1) November-December 2015

Identify a number of areas to be considered for a vision setting process and create a survey instrument - Completed.

Conduct a SAB/MAB survey to poll their opinions on future directions and priorities (Vision Setting Survey) - Completed.

Start the user survey planning process - Completed.

Hire a part-time person (or assign this role to an existing staff) to work with us in coordinating the planning activities - Completed.

(2) January-February 2016

Analyze and report findings of the vision setting exercise - Completed.

Start the 2016 annual fund raising campaign with the goal of increasing annual contributions (arXiv MAB subgroup will come up with recommendations to increase annual membership revenue through a platinum membership model) - Completed.

Plan and test a user survey based on the initial findings of the vision-setting exercise, taking into consideration different segments: users as readers, users as submitter, users as moderators - Completed.

Receive recommendations from the MAB IT subgroup about the goals and desired outcomes of a workshop with repository experts to consider arXiv's current infrastructure and workflows & put in place a plan for hosting a workshop (Planning Meeting I) - Completed.

(3) March-May 2016

Conduct the user survey - Completed.

Host IT external experts/consultants to advise us on future directions (Planning Meeting I)-Completed Additional Information.

Start analyzing and reporting the findings from the vision setting process, user study, and the IT workshop-Completed.

(4) June-July 2016

Prepare a report summarizing the key findings of the user survey - Completed.

Based on the findings of the vision setting process, user study, and the IT workshop, start identifying priorities and funding sources to support next-gen arXiv development efforts.

Repeat the online fund raising campaign (Give button) to generate additional funds.

Based on the findings of the user study and the IT workshop, consider hosting a meeting with a small group (10-12) to discuss our preliminary findings and consider the future of arXiv and the evolving nature of scientific communication (Planning Meeting II: arXiv@25) - Decided not hold such a meeting as the team has already gathered signfichat input.

Design and implement a survey targeting arXiv's moderators (new addition to the strategy, June 10, 2016).

(5) August-September 2016

Continue to identify funding sources, make initial contacts.

Plan and hold SAB/MAB annual meeting to review findings and nascent work plans & develop strategies for moving us forward.

(6) October-December 2016

Vision-grant writing and creating work plans (our hope is that by then through the planning activities during the last several months, we'll already have ideas for funding sources and even might have done some initial work).

# 9  arXiv User Survey Report

## 9.1  Executive Summary

As part of its 25th anniversary vision-setting process, the arXiv team at Cornell University Library conducted a user survey in April 2016 to seek input from the global user community about the current services and future directions. We were heartened to receive 36,000 responses, representing arXiv's diverse community (See Appendix A). The prevailing message is that users are happy with the service as it currently stands. 95% of survey respondents said that they are very satisfied or satisfied with arXiv. Furthermore, 72% of respondents indicated that arXiv should focus on its main purpose, which is to quickly make available scientific papers, and this will be enough to sustain the value of arXiv in the future. This theme was pervasively reflected in the open text comments. A significant number of respondents suggested keeping to the core mission and enabling arXiv's partners and related service providers to continue to build new services and innovations on top of arXiv.

Many of the comments reflected deep satisfaction with and gratitude for arXiv. Several users referred to the significance of the service for their personal career development and expressed thanks for its continued existence; for example, a typical comment was: "Thanks for the hard work of many people over the years. My work life would be very different without your efforts." arXiv also received many plaudits for advancing the dissemination of research through the open-access system. One user referred to the service as "a beacon for scientific communication." Several commenters expressed how crucial arXiv has been for them personally in enabling them to quickly access the latest research in their field. There was an overall perception that arXiv was an important leader in the development of alternatives to traditional publishing. Independent researchers who are unaffiliated with large institutions and who might otherwise have delayed access to papers particularly emphasized the importance of arXiv for their work.

The combination of multiple choice responses (see Appendix B) and the extensive and thoughtful open text comments pinpointed areas that need to be upgraded and enhanced. Improving the search function emerged as a top priority as the users expressed a great deal of frustration with the limited search capabilities currently available, especially in author searches. Providing better support for submitting and linking research data, code, slides and other materials associated with papers emerged as another important service to expand. Regardless of their subject area, users were in agreement about the importance of continuing to implement quality control measures, such as checking for text overlap, correct classification of submissions, rejection of papers without much scientific value, and asking authors to fix format-related problems. Several users commented on the need to randomize the order of new papers in announcements and mailings. There were several useful remarks about the need to improve the endorsement system and provide more information about the moderation process and policies.

In regard to arXiv's role in scientific publishing, some users encouraged the arXiv team to think boldly and further advance open access (and new forms of publishing) by adding features such as peer review and encouraging overlay journals. On the other hand, many users strongly emphasized the importance of sticking to the main mission and not getting side-tracked into formal publishing. There was a similar divergence of opinion about encouraging an open review process by adding rating and annotation features. When it comes to adding new features to arXiv to facilitate open science, the prevailing opinion was that any such features need to be implemented very carefully and systematically, and without jeopardizing arXiv's core values.

While many respondents took the time to suggest future enhancements or the finessing of current services, several users were strident in their opposition to any changes. Throughout all of the suggestions and regardless of the topic, commenters unanimously urged vigilance when approaching any changes and cautioned against turning arXiv into a "social media" style platform. The feeling is that arXiv as it exists is working well and while there are some areas for improvement, too much change could potentially weaken the effectiveness and overall mission of arXiv.

## 9.2  Key Findings

### 9.2.1  Improving the Current arXiv Services

When asked about the importance of improving a specific range of services, more than 70% of respondents said that improving search functions to allow more refined results was very important/important across all groups by years of use, age groups, number of articles published, country groups, and subject areas. Many commenters requested enhanced functions such as author search, date-limited searching, and searching non-English languages. Search was equally problematic regardless of whether the user searched for a known paper, was browsing a subject category, or looking for specific authors.

A series of questions asked users about improving the submission process specifically with (1) support for submitting research data, code, slides and other materials; (2) improving support for linking research data, code, slides, etc., with a paper; and (3) updating the TeX engine and various other enhancements. Support About 40% of respondents rated each one as very important/important. The open text responses also displayed considerable interest in better support for supplemental materials, although respondents disagreed as to whether they should be hosted by arXiv or another party. Many respondents are supportive of integrating or linking to other services (especially GitHub), while a significant number of respondents also indicated doubts about long-term availability and link rot for content not hosted within arXiv. Some expressed concerns regarding the resources required for arXiv to improve this. There was some interest in including the data underlying figures in arXiv papers.

Among other services and improvements recommended by respondents were:

Consistent inclusion of information and links about the published versions of the papers.

More refined options for alerting, both email and RSS. Several respondents specifically requested email alerts for works by a particular author, and there was some interest in HTML-formatted email with live links.

Updating and keeping current arXiv's TeX engine and provide TeX templates or style files to make submission easier.

Linking papers to each other via citations and actionable links in bibliographies.

Ability to submit a PDF, an increase in the file size limit (often with specific request to link to figures), and the ability to upload multiple files at once.

Allowing submission directly from authoring platforms (such as Overleaf or Authorea).

Providing use statistics such as paper downloads and views.

A much larger percentage of recent arXiv users (five years or less) selected the "no opinion" option about current service upgrades. For all the questions in this category, the same trend is visible: a higher percentage of recent users expressed that they had no opinion and this percentage of respondents decreased with each level of increase in years of use. Interestingly, this same trend is not visible by age group; i.e. our data do not show that a higher percent of younger users have no opinion.

## 9.2.2 Importance of Quality Control Measures

arXiv's users were asked a series of questions regarding quality-control measures. Based on the 26,430 responses to specific controls, the most important of these (ranked very important/important) were:

| | |
|---|---|
| Check papers for text overlap, i.e., plagiarism | 77% |
| Make sure submissions are correctly classified | 64% |
| Reject papers with no scientific value | 60% |
| Reject papers with self-plagiarism | 58% |

A large percentage of all demographic groups found checking for plagiarism to be important and a slightly smaller group found checking for self-plagiarism as important. There was no discernible difference across demographic groups for the other measures. Similarly, self-plagiarism was also mentioned as another area for improvement. Some noted that context is the key; for example, conference papers are a common and typical area where self-plagiarism could occur in an otherwise scientifically sound submission.

Several respondents said they were unaware of precisely what quality-control measures were already in place, and felt that the process is too opaque. Others acknowledged the difficult balance between rejecting papers that are clearly unworthy—"crackpot"—and rejecting papers for other, perhaps less obvious, and anonymized reasons. However, even in the face of such criticisms there was a strong thread of satisfaction with arXiv's current quality-control process and users cautioned against going too far in the other direction.

Some users would prefer that arXiv embrace a more open peer review and/or moderation process, while others were adamant that current controls allow arXiv the freedom and speed of access that is otherwise unobtainable through traditional publishing.

Overall, the feeling was that quality control matters but user comments varied greatly in relation to how arXiv could practically achieve these goals. As one respondent wrote, "Judgment about quality control is a very relative issue".

### 9.2.3 Adding New Subject Categories

73% of the respondents were not interested in seeing new subject categories added to arXiv. 26% of respondents would like to see new subject categories added and suggested chemistry (881), engineering (483), biology (429), economics (248), philosophy (220), and social sciences (106). There were also several smaller categories such as Machine Learning (82 responses) and Artificial Intelligence (27 responses).

A frequently repeated theme was that arXiv does not need to focus particularly on additional subjects but instead should focus on the refinement and addition of subfields and subcategories, especially in High Energy Physics Theory as well as Mathematics.

### 9.2.4 Developing New Services

Users were asked to rate a range of proposed new services for arXiv. In the ranked responses, more than 63% of users rated adding direct links to papers in the references (reference extraction) as very important/important. Citation export in formats such as BibTex, RIS was rated as very important/important by more than 57% of users, and extraction for the BibTeX entry for the arXiv citation was similarly rated by more than 55% of respondents. Citation analysis tools in general were ranked as very important/important by almost 53% of respondents.

In the open text comments, opinions were divided on the need for enhanced citation-analysis capabilities. While users were generally in favor of citation tools many of the same users noted that other systems are already doing this, and that this was sufficient for their needs.

In the multiple choice survey responses the option to "offer a rating system so readers can recommend arXiv papers that they find valuable" was closely split between very important/important (36%) and not important/should not be doing this (36%). This matches the way the comments were closely split between those in favor and those less certain. Also, it was found valuable by 50% of recent users as compared with 28% of seasoned users. In addition, a larger percentage of younger users find it important (42% of those under 30 years), as compared to 28% of those 60 and above. Opinions were divided in the open text comments but overall the respondents were hesitant about the idea. Some users liked the rating feature "in an ideal world" setting, but did not think it was appropriate for arXiv; others expressed concern that it would dilute the mission of arXiv, or simply appears unfeasible in arXiv's current incarnation. However, even users directly in favor of a rating system raised issues about whether it would be open to the public, rated by peers, anonymous, etc. Several respondents stressed that such a feature would need to be implemented very carefully.

Like the question about offering a rating system, the idea of adding an annotation feature to allow readers to comment on papers was almost evenly split, with 34.89% of users ranking it as very important/important and 34.08% as not important/should not be doing this. In the open text responses, the trend opposed the idea and some of the responses reflected strongly negative feelings. Those in favor or open to the idea of a commenting system often added a caveat and in general there was a sense of caution even for those responding positively. A common theme of concern was that a moderated system and verifiable accounts would be necessary to prevent a free-for-all. Unlike the question about offering a rating system, there were no discernible differences in opinion based on different demographic characteristics.

### 9.2.5  Finding arXiv Papers

The vast majority of arXiv's users access the papers directly from the homepage (79%), followed by using Google to search (50%) and Google Scholar (35%).

Once on the homepage, reactions were mixed regarding the ease of use and navigation. 32% rated this as easy, but only 25% find it somewhat easy and 21.6% rated it somewhat difficult to use.

To discover content, 63% of users go to the link for new or recent under a particular category and equally 63% of users use arXiv's search engine and enter a specific arXiv ID, author name or search term. A small number of users, 14%, rely on the daily mailing list and then look for a particular article in the search field.

In the open text comments, opinion was divided about the user interface. The majority of respondents disliked the outdated style, but a definite subgroup appreciated the interface's simplicity, which these users feel helps arXiv efficiently carry out its mission. The main issues mentioned aside from the homepage's look were the number of links, layout and finding submission information.

The lack of hierarchy in organization was found challenging to understanding arXiv's navigation.

Requests for enhancements related to UX included greater personalization of arXiv for readers; for example, the ability to "favorite" papers, curate a personal library, and see recommendations when users visit the site. Other users mentioned the development of APIs to further facilitate the development of overlay journals. Some users also suggested the development of a mobile-friendly version.

Many commenters either described how they rely on other services to interact with arXiv content (site-specific searches, ADS, INSPIRE) or recommended features based on their experience with other information systems. Among those frequently praised were ADS, INSPIRE, Google Scholar, gitxiv.com and arxiv-sanity.com.

About arXiv: arXiv, an open-access scientific digital archive, is funded by the Simons Foundation, Cornell University Library, and about 190 member libraries from all around the world. The site is collaboratively governed and supported by the research communities and institutions that benefit from it most directly, ensuring a transparent and sustainable resource. It is a moderated scholarly communication forum informed and guided by scientists and the scientific cultures it serves. As of June 2016, arXiv contains more than 1,110,000 e-prints. In 2015, the repository saw 105,000 new submissions and close to 139 million downloads from all over the world.

## Appendix A: Demographics of Respondents

I use arXiv in the following ways: (Please choose all that apply)

| Answer | Ratio | Count |
|---|---|---|
| I am an arXiv reader | 93% | 31,862 |
| I am an arXiv author | 53% | 18,270 |
| I am an arXiv submitter | 50% | 17,189 |
| I am an arXiv (other type of user): Please describe | 2% | 845 |

The number of articles I have published/submitted on arXiv is:

| Answer | Ratio | Count |
|---|---|---|
| 1 article | 11.99% | 2,570 |
| 2 articles | 8.96% | 1,920 |
| 3-4 articles | 15.19% | 3,254 |
| 5-10 articles | 23.06% | 4,941 |
| More than 10 articles | 40.80% | 8,743 |
| Total | 100% | 21,428 |

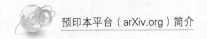

My current occupation is: (Please choose ALL that apply)

| Answer | Ratio | Count |
|---|---|---|
| I am an academic faculty member (professor) at a college or university | 27% | 8,868 |
| I am an academic staff member (researcher or postdoc) at a college or university | 22% | 7,207 |
| I am a researcher at a non-profit or governmental agency | 8% | 2,707 |
| I am a Masters/Ph.D. student | 30% | 9,890 |
| I am an undergraduate student | 5% | 1,514 |
| I am (please describe) | 13% | 4,353 |

13% of respondents (4,353) indicated a different occupation category. The top ones included researchers at a company or industry (900), engineer (515), and retired individuals (478). There were also respondents who described themselves as science writers, editors, or freelance editors. Other response types included data scientist, self-described amateur researchers, self-described laypeople, unemployed, teachers, and the generally curious (e.g. "a man doing research as hobby").

As a user, my main subject area of interest in arXiv is: (please choose all that apply)

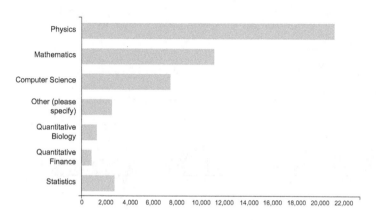

Almost 2,000 respondents checked the Other option to specify their main area of interest. The top categories were astrophysics (726) and astronomy (653).

I have been using arXiv for:

| Answer | Ratio | Count |
|---|---|---|
| 0-2 years | 19.54% | 6,470 |
| 3-5 years | 28.96% | 9,592 |
| 6-10 years | 25.44% | 8425 |
| 11 or more years | 26.06% | 8,632 |
| Total | 100% | 33,119 |

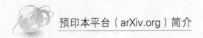

预印本平台（arXiv.org）简介

My age is:

| Answer | Ratio | Count |
|---|---|---|
| younger than 30 years | 37.42% | 12,364 |
| 30-39 years | 31.27% | 10,332 |
| 40-49 years | 13.76% | 4,545 |
| 50-59 years | 9.30% | 3,073 |
| 60-69 years | 5.77% | 1,908 |
| 70 years and over | 2.47% | 817 |
| Total | 100% | 33,039 |

My main place of work is located in:

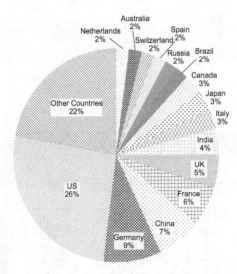

Other Countries: 1% or less representation each from 113 countries

## Appendix B: Opinions on arXiv's Current Services & Future Directions

How important is it to improve on the following CURRENT arXiv services?

| Question | Very important & important | Somewhat important | Not important & should not be doing this | No opinion |
|---|---|---|---|---|
| Improve search functions to allow more refined results (e.g. narrow down results by additional search terms, filter by publication year or institutional affiliation, etc.) | 70.38% | 19.34% | 6.14% | 4.13% |
| Improve support for submitting research data, code, slides, and other materials associated with a paper (e.g. I want to be able to upload my datasets/ machine- readable tables with my article) | 41.95% | 22.64% | 14.03% | 21.37% |

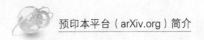

(Continued)

| Question | Very important & important | Somewhat important | Not important & should not be doing this | No opinion |
|---|---|---|---|---|
| Improve support for linking research data, code, slides, and other materials associated with a paper (e.g. I want to be able to link to my slides on SlideShare) | 40.65% | 25.20% | 17.70% | 16.45% |
| Improve support for submitting research papers by updating the TeX engine | 39.36% | 23.17% | 16.71% | 20.76% |
| Improve the email alert system so that readers can customize their settings and choose to receive alerts about specific sub-topics | 37.85% | 26.48% | 20.25% | 15.42% |
| Improve the trackback mechanism (linking papers back to blogs and commentaries that cite thos papers) | 36.52% | 29.50% | 20.30% | 13.67% |
| Simplify the submission process by providing clearer instructions and simpler language | 32.45% | 22.55% | 25.20% | 19.80% |

How important is it to develop the following NEW arXiv services?

| Question | Very important & important | Somewhat important | Not important & should not be doing this | No opinion |
|---|---|---|---|---|
| Add direct links to papers in the references (support reference extraction) | 63.04% | 26.89% | 5.78% | 4.29% |
| Offer citation export in formats such as BibTeX, RIS, etc. | 57.68% | 23% | 10.95% | 8.37% |
| Enable extraction for the BibTeX entry for the arXiv citation | 55.54% | 23.82% | 9.72% | 10.91% |
| Provide Citation Analysis tools (examining the frequency and pattern of a paper's citation) | 52.95% | 27.08% | 14.28% | 5.69% |

(Continued)

| Question | Very important & important | Somewhat important | Not important & should not be doing this | No opinion |
|---|---|---|---|---|
| Support compliance with public/open access mandates (funding agency policies that require research results to be made public) by allowing final versions of papers to be submitted with information such as funding sources and grant numbers | 42.06% | 26.21% | 13.68% | 18.05% |
| Enable submitting an article to a journal at the same time as it is uploaded to arXiv | 39.28% | 23.09% | 25.23% | 12.40% |
| Offer a rating system so readers can recommend arXiv papers that they find valuable | 36.28% | 21.76% | 35.56% | 6.40% |

# 9 arXiv User Survey Report

(Continued)

| Question | Very important & important | Somewhat important | Not important & should not be doing this | No opinion |
|---|---|---|---|---|
| Enable linkages (interoperability) with other repositories (e.g.,run by libraries), so that a paper accepted by arXiv is accepted at the same time by the other repositories | 35.25% | 28.14% | 17.25% | 19.36% |
| Develop an annotation feature which will allow readers to comment on papers | 34.89% | 23.62% | 34.08% | 7.41% |

Where do you go to find arXiv papers? Please choose all that apply.

| Answer | Ratio | Count |
|---|---|---|
| Go directly to arXiv.org (arXiv homepage) | 79% | 22,804 |
| ADS | 14% | 4,144 |
| Inspire | 13% | 3,773 |
| Google Scholar | 35% | 10,016 |
| Google search engine | 50% | 14,440 |
| arXiv email alerts | 14% | 4,086 |

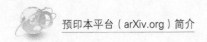
预印本平台（arXiv.org）简介

(Continued)

| Answer | Ratio | Count |
|---|---|---|
| Other search engines | 5% | 1,402 |
| Subject gateways for arXiv, such as the Math Front | 4% | 1,203 |
| Other (please specify) | 9% | 2,662 |

If you have used the arXiv homepage for finding papers, how easy is it to navigate?

| Answer | Ratio | Count |
|---|---|---|
| Very easy | 14.85% | 3,916 |
| Easy | 32.05% | 8,450 |
| Somewhat easy | 25.20% | 6,644 |
| Somewhat difficult | 21.60% | 5,696 |
| Difficult | 5.02% | 1,324 |
| Very difficult | 1.27% | 336 |
| Total | 100% | 26,366 |

If you have used the arXiv homepage, how do you usually navigate our main page? Please choose all that apply.

| Answer | Ratio | Count |
|---|---|---|
| Go to link "new" or "recent" under a particular category | 63% | 16,503 |
| Use arXiv search engine and enter a specific arXiv-id, author name, or search term | 63% | 16,478 |

(Continued)

| Answer | Ratio | Count |
|---|---|---|
| Receive daily mailing list, and then look for a particular article on the search field | 14% | 3,692 |
| Other, please explain | 3% | 853 |

How important are the following CURRENT quality control measures?

| Question | Very important & important | Somewhat important | Not important & should not be doing this | No opinion |
|---|---|---|---|---|
| arXiv checks papers for text overlap: an author's use of too much identical text from other authors' papers, without making it clear that the text is not their own material, i.e. "plagiarism" | 77.41% | 14.66% | 4.96% | 2.96% |
| arXiv makes sure submissions are correctly classified (the subject categories are included on the arXiv homepage) | 64.38% | 25.32% | 7.01% | 3.29% |
| arXiv keeps out (rejects) papers that don't have much scientific value | 60.02% | 19.14% | 15.49% | 5.35% |

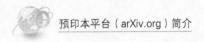

(Continued)

| Question | Very important & important | Somewhat important | Not important & should not be doing this | No opinion |
|---|---|---|---|---|
| arXiv checks papers for too much text re-use from an author's earlier works, i.e. "self-plagiarism" (reuse of identical content from one's own published work without citing) | 57.77% | 24.64% | 14.08% | 3.51% |
| arXiv checks papers for format-related problems (line numbers in text, missing references, oversize submissions, etc.) and asks authors to fix them before they are announced | 55.00% | 29.83% | 11.51% | 3.66% |
| arXiv moderates the scientific content of trackback (links to blogs and commentaries) before permitting the link to be added | 39.60% | 26.31% | 17.59% | 16.50% |

## Overall, how satisfied are you with arXiv?

| Answer | Ratio | Count |
|---|---|---|
| Very satisfied | 52.92% | 14,770 |
| Satisfied | 42.43% | 11,841 |
| Somewhat satisfied | 3.55% | 990 |
| Somewhat dissatisfied | 0.54% | 150 |
| Very dissatisfied | 0.15% | 42 |
| No opinion | 0.42% | 116 |
| Total | 100% | 27,909 |

## Which of the following BEST describes your opinion of how arXiv needs to move forward?

| Answer | Ratio | Count |
|---|---|---|
| arXiv should focus on its main purpose, which is to quickly make available scientific papers. This will be enough to hold up the value of arXiv in the future | 71.94% | 19,865 |
| arXiv should expand its main mission, and spend more time and resources to provide new services. This is necessary to hold up the value of arXiv in the future | 19.59% | 5,410 |
| No opinion | 8.47% | 2,340 |
| Total | 100% | 27,615 |

# 1 arXiv 的一般性问题

## 1.1 arXiv 是什么

arXiv.org 是公认最成功的开放获取数字系统之一。1991年8月，Paul Ginsparg 创建了 arXiv，将它作为物理学领域的预印本存储知识库。2001年，这一平台转移到康奈尔大学图书馆。如今，它改变了物理学多个领域的学术交流方式，并且在数学、计算机科学、生物计量学、金融计量学和统计学等领域发挥着越来越突出的作用。它被嵌入到这些学科的研究工作中并让科学成果的快速传播成为可能。通过供全世界的科研人员开放获取，arXiv 使科学变得更加自由开放。

截至2012年8月，arXiv 已包含超过 770 000 份电子文档。2011年，这一知识库收到了 76 578 份提交文档和来自世界各地的将近 5000 万次下载。arXiv 是国际性的，在 9 个国家都有镜像网站，并与美国和国外专业社团及其他国际机构都有合作。

## 1.2 arXiv 的主要准则

arXiv 为全世界的作者和研究人员提供了一个科学研究的开放获取知识库。这是一个有组织的学术交流论坛，由科学家和它所服务的科学文化通知和引导。arXiv.org 对所有终端用户都是免费的，个人研究人员可以将他们的研究内容免费存放在 arXiv 中。

## 1.3 提交文章的合适方法

我们期待提交给 arXiv 的材料是有益的、相关的并且对这些学科是有价值的。arXiv 保留对任何提交材料拒绝或进行重新分类的权利。专家们审查提交的文档来确定它们是否是当前热点且与科研贡献相关，同时符合学术交流的准则（以传统期刊为例）。提交材料和审核政策请参考 http://arxiv.org/help。

# 2　可持续发展倡议

## 2.1　倡议的目的

自 2010 年起,康奈尔大学图书馆的可持续发展计划倡议就已经把目标设定为减少 arXiv 的经济负担和对单一机构的依赖,取而代之的是创建一个拥有广泛基础和社群支持的资源。为了保证一个开放获取的学术知识库 arXiv 的可持续发展,管理者必须在持续提高其价值的基础上,满足所有用户社群的需求,以及覆盖相关的运营成本。康奈尔倡议旨在努力加强 arXiv 基础设施的技术、服务与政策。

## 2.2　结果

作为临时性政策,康奈尔大学图书馆首先建立了一套自愿贡献的制度,邀请了全世界范围内 200 所图书馆和研究实验室担任 arXiv 最初最重要的机构用户。在这一 2010—2012 计划进程中,康奈尔大学图书馆寻求来自广泛的利益相关者的合作管理,社群支持的资源。规划工作还包括与几个出版商和社会团体的代表会议,讨论合作的可行性和可取性,这将提高科学文献的互联、互操作性和支持科研生命周期的服务。

可持续发展计划进程的目的在于调查如何将收入模式多元化，确保 arXiv 符合一系列的管理准则，并且提供透明化的、可靠的社群支持服务。arXiv 的会员模式是 3 年计划进程的关键成果，这一计划进程是由西蒙斯基金资助的。这一计划活动的背景信息可在 http://arxiv.org/help/support 上获取。

# 3　arXiv 的会员模式

## 3.1　arXiv 会员项目的目标

为了协助 arXiv 的支持和管理，会员项目的目的是与代表 arXiv 重要机构用户的全世界的图书馆和研究实验室建立密切关系。每个成员机构承诺 5 年初始资金支持 arXiv。arXiv 可持续发展计划建立在 arXiv 的运营标准之上，提出了一个可以创造收入的经营模式。

## 3.2　arXiv 的收入模式

康奈尔大学图书馆、西蒙斯资助机构和全球机构成员支持 arXiv 的财务。2013—2017 年的财务模型涉及 3 个收入来源：

● 康奈尔大学图书馆提供每年 75 000 美元的现金补贴支持 arXiv 的运营成本和一些间接成本，这些目前占总营业费用的 37%。

● 西蒙斯基金会每年贡献 50 000 美元为了康奈尔大学图书馆对 arXiv 的管理。除此之外，这一资助机构每年通过 arXiv 会员费而匹配了 300 000 美元。

- 每个成员机构承诺 5 年初始资金支持 arXiv。根据机构用户的分级，每年的费用有 4 个等级，从 1500 美元到 3000 美元不等。康奈尔大学图书馆的目标是每年通过 126 个机构筹集 300 000 美元的费用。

其中，西蒙斯基金会每年匹配资金的目的是通过降低 arXiv 会员费鼓励长期的社群支持，使更多样的机构有能力参与。这种支持有助于确保可持续发展的 arXiv 的最终责任，使得研究团体和机构可以从服务中直接受益。

## 3.3 arXiv 的运营成本

arXiv 2013—2017 年的计划运营成本平均每年为 826 000 美元，包括间接的费用。2012—2017 年的运营预算计划包括上面提到的 4 个主要的收入来源：康奈尔大学图书馆每年 75 000 美元的资助，包括间接的费用；西蒙斯基金会每年 50 000 美元的资助；每年来自成员机构的会员费；以及每年 300 000 美元西蒙斯基金会基于会员费的收入。

其中，预算中包含应急基金，arXiv 应急基金支持紧急开支，以确保一个健全的运行模式。如今，arXiv 已拥有大约 125 000 美元的应急基金，来自过去 2 年由于突发的工作人员空缺和其他的储蓄积累，如向虚拟服务过渡等。这些资金的使用准则需与会员咨询委员会讨论。

## 3.4 机构会员的费用

会员费用基于机构的等级来计算,级别的分类通过文章的下载量而定。用户数据和机构等级每年都要计算。2009—2011 年用户数据前 200 名机构用户在 arXiv 可在线获取。为了鼓励大家参与 arXiv 会员,费用将随着会员数目的增加而减少。

| 用户等级 | 会员年费 |
| --- | --- |
| 1～50 | $3000 |
| 51～100 | $2500 |
| 101～150 | $2000 |
| 151+ | $1500 |

## 3.5 成为会员的好处

参与组织所独享的好处有:

● 通过会员咨询委员会参与 arXiv 的管理,它能渗入项目的优先次序、新的服务产品、财务规划、自由支配资金的使用、未来的技术发展和政策决定;

● 提高机构使用数据的方式;

● 在经济支持上获得公共的认可。

一些其他的好处正在发展中,包括自动提交 arXiv 文档到作者主管机构的机构知识库中,以及创建一个会员的门户为参与机构提供及时的信息。

## 3.6 重点关注前 200 名机构用户

使用是通过文章的下载量计算的。康奈尔大学图书馆关注前 200 名机构，因为它们占了下载量的 75%。这不是一个迫使任何人支持 arXiv 的内容获取而强制性收费的资助模式，提交者和使用者访问 arXiv 都是免费的。arXiv 同时乐于接受和认可来自其他图书馆和研究实验室的支持。

## 3.7 成员机构的认可方式

可持续发展信息，包括一系列的 arXiv 支持者，在主菜单的帮助选项下可用 http://arxiv.org/help/support。通过 IP 识别，这些机构的用户可以看到认可的标示，以表示得到他们机构的支持。康奈尔大学图书馆正计划将这些信息更加突出显示给 arXiv 的用户，具体地传达参与者对机构科学家的贡献。这对提高对运行知识库的资源需求的认识和成员图书馆在确保 arXiv 的未来的角色方面至关重要。

## 3.8 模式是否公平

arXiv 是物理学和数学核心领域作者和读者的主要目的网站。arXiv 的可持续发展应被看作是文化嵌入资源，它为全球科学研究者网络提供明确的价值共享投入。任何一个自愿贡献的系统都存在着"搭便车"的情况，但是 arXiv 非常符合成本效益，所以与订阅服务相比，重要用户机构的巨大贡献将支持持续的开放获取，物超所值。arXiv 会员模型为参与的组织根据他们对 arXiv 的贡献提供独家好处。

## 3.9 会员承诺

### 3.9.1 申请会员的资格

arXiv 的会员和会员咨询委员会的代表位置是为服务提供财务支持的图书馆、研究机构、实验室和基金会保留的。

### 3.9.2 提名候选人的流程

这一信息在 2012 年 9 月增加至网站 http://arxiv.org/help/support。

### 3.9.3 承诺协议

康奈尔大学图书馆设想了一个 5 年承诺作为一个支持 2013—2017 年 arXiv 的良好愿景。它不被看作是一个一次性付款或具有法律约束力的合同。

### 3.9.4 开具发票的时间线

机构于 2013 年 3 月 31 日之前开具发票。时间线是灵活的，而且对于曾经的资助者，康奈尔大学图书馆会保留当前的发票时间表。获取发票日期请联系我们：support@

arxiv.org。

排名表前 200 名以外的机构申请成为支持性会员，同样请发邮件给我们：support@arxiv.org。

### 3.9.5 机构提供 IP 地址

会员将在 arXiv.org 网站上公开。机构提供 IP 地址信息，康奈尔大学图书馆可以在 arXiv 标志右上方自定义信息："我们非常感谢_____的支持。"

### 3.9.6 查找信息

arXiv 的 2009 年、2010 年和 2011 年的机构数据可以在线获得。如果您的机构未在列表之上，请给我们发邮件获取使用数据：support@arxiv.org。

我们经常收到关于提供基于提交的统计数据，并比较他们与我们目前的机构下载统计数据的可行性。目前，arXiv 的作者元数据不具有充分的一致性来支持任何系统做分析。我们已经进行了单月提交数据的手动分析，样本的统计结果显示基于提交和下载的数据在机构排名的特征上表现出了相似的特征。我们继续完善我们的研究结果，并期待这些信息可以用于经营规划用途的想法。随着时间推移，arXiv 想要进行元数据的修复工作从而提高现有文稿的作者数据。这将有助于更好地进行作者链接，改进所有权的说明及支持作者身份链接使用 ORCID 作者标识符奠定基础。

# 4 arXiv 的管理模式

## 4.1 管理模式的运作

在会员咨询委员会和科学咨询委员会的指导下，康奈尔大学图书馆拥有全部的 arXiv 运营和发展的行政和财务责任。康奈尔大学图书馆管理提交和用户支持的审核，包括政策和程序的发展和实施、运行 arXiv 技术基础设施、承担存储的责任并确保长期获取、负责 arXiv 的镜像网站，并与相关项目保持合作，从而通过互操作性和合作共享提高对科研团体的服务。

其中，科学咨询委员会是由 arXiv 涵盖学科的科学家和研究人员组成的。它提供了建议和有关知识库知识的监督指导，并特别关注 arXiv 政策和稳定系统的运行。科学咨询委员会审查和修改 arXiv 的存储准则和标准；为 arXiv 的存储提出新的学科或学科领域；提供会员咨询委员会专家们对 arXiv 发展计划提出的反馈；制定关于开发项目的建议（尤其是关于提升系统以支持提交和适度处理的建议）。

会员咨询委员会由从 arXiv 的会员中选举出来的代表组成，作为一个咨询机构。委员会代表着参与机构的利益并向康奈尔大学图书馆提供与知识库管理和发展、标准的实施、互操作、优先发展、经营计划、财务计划及推广和宣传相关的建议和问题。arXiv 的会员及会员咨询委员会的代表是为提供给 arXiv 金融支持服务的图书馆、研究机构、实验室和基金会保留的。会员咨询委员会代表的选举投票对所有选举机构开放，每个机构都有一票。会员咨询委员会章程指定了会员委员会的业务，包括组成、成员的任命方法和任期。

arXiv 已经拥有一个科学咨询委员会，是由 Paul Ginsparg 在几年前建立的。为了给组织创造一个规章制度，康奈尔大学的 arXiv 团队正在收集科学咨询委员会组成、条款和委员会会员选举方面的细节信息。2012 年 4 月，康奈尔大学图书馆成立了一个临时的会员咨询委员会，审核和批准会员咨询委员会的章程，制订其流程，包括会员任命方法、任期等。这一临时的组织同时旨在明确 arXiv 成功的方法和新型的管理模式，它将测试潜在的治理方案，以更好地了解康奈尔大学图书馆、会员咨询委员会和科学咨询委员会之间的互动。这将确定各自的责任，细化 arXiv 的治理准则。

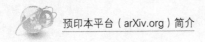

## 4.2 新管理模式的启动时间

从 2012 年起的 3 年的可持续性规划阶段向康奈尔大学图书馆、会员咨询委员会和科学咨询委员会角色清晰的阶段过渡时期。

# 5 职员和费用

## 5.1 arXiv 2013—2017 年的职员预测和支出预测

职员预测已包含在经营模型文件中。

arXiv 的经费是以一个康奈尔大学图书馆资金账户的结构存在，目的是遵循康奈尔的财务政策并提供审计跟踪。arXiv 的支出预测在经营模型文件中有描述。

## 5.2 附加收入来源

在调查经营模式的过程中，康奈尔大学图书馆考虑了很多选择，排除了论文处理费用和提交费用。无限制的提交和使用是 arXiv 的创始原则之一。arXiv 的咨询委员会将考虑增加其他收入来源，例如，在 arXiv 网页上增加"打赏"按键，从而鼓励科学家和其他兴趣团体进行捐赠。

## 5.3 arXiv 和 SCOAP3 之间的关系

arXiv 和 SCOAP3（国际高能物理开放出版资助联盟）在范围和功能上都是互补的。arXiv 是一个学术交流论坛，在广泛的领域，包括物理学、数学、计算机科学、定量生物学、定量金融和统计研究人员的快速开放的访问传播方面服务研究人员的需要。SCOAP3 解决了高能物理领域同行评议出版物的可持续开放获取问题，高能物理是 arXiv 支持的众多数据集之一（提交的论文量占 arXiv 论文总量的 20% 以上）。这些出版物比 arXiv 预印本文档晚一年出版，对于这些科研团体非常重要，因为其将作为供高校和资助机构问责制使用的最终版本，而学术话语权在 arXiv 手中（详见 http://arxiv.org/abs/0906.5418）。

然而 SCOAP3 与出版商合作提供正式出版物的开放获取，arXiv 是一种学术交流论坛，由它服务的科学家和科学文化主导。专家负责审查意见书，确认它是否是当前热点且符合科研团体的兴趣，遵循公认的学术交流标准，并分类在相应学科类别中。因此，arXiv 分类对于区分 SCOAP3 赞助的出版物方面是一个有价值的关键因素。随着 SCOAP3 计划的不断发展，arXiv 和 SCOAP3 完全致力于进一步讨论在开放获取科学交流领域的共享机会，利用 2 个团队在其他项目上强大的合作纽带作用。在这个事情上，arXiv 的财务计划并不假定任何直接从 SCOAP3 项目而来的资源。

# 6 arXiv 在康奈尔大学图书馆

## 6.1 人员配备

arXiv 的直接运营包括大约 6 名全职员工（FTE）。超过半数以上的员工是参与支持用户和管理系统的。其他正式员工提供项目和系统支持，还有少数属于管理层。其他的管理和业务支持属于运营管理，例如项目总监对 arXiv 的贡献管理。

## 6.2 接收研究数据

2011 年，arXiv 为 arXiv 文章的相关数据开发了一个试点数据上传接口。数据的提交是通过 arXiv 提交界面的拓展界面完成的。当文章发布并存储在 arXiv 时，数据自动存入数据库并链接到文章（详见：http://arxiv.org/help/data_conservancy）。这是一个前导计划，在 2012 年年底重新评估与数据库的合作。少量的数据、程序代码等可能会与 arXiv 文章一起作为辅助文件（详见：http://arxiv.org/help/ancillary_files）。

## 6.3 保存策略

数据保存是指一系列支持比特流长期维护的管理活动。这些活动确保了数据对象是可用的（完整和可读的），保留全部的真实性、精度和文章（以及其他相关材料）存入时不可缺少的功能。被 arXiv 接纳的格式是基于它们的档案价值（TeX/LaTeX、PDF、HTML）和灵活监控处理所有源文件的能力。底层的比特流受康奈尔大学校园的标准备份程序保护。纽约市异地备份设施提供地理冗余。全部的内容都会复制到 arXiv 在全世界的镜像网站，其他的内容保存在洛斯阿拉莫斯国家实验室中。康奈尔大学图书馆拥有一个档案知识库支持机构关键资源的保存，包括 arXiv。我们预期在该知识库存储所有的 arXiv 文件，包括来源和过程形式。以后逐步会添加新的材料。我们期望康奈尔大学图书馆承担 arXiv 的数据保存成本，作为图书馆系统研发的档案基础设施。

## 6.4 学科扩展

由于创造和维护新的学科领域需要付出大量的组织和管理努力，因此我们已经采纳了扩展学科的谨慎的方法。增加一个新的学科领域包含挖掘用户群，利用特征有关的学科领域，建立必要的咨询委员会，并招募管理者。同时，尽管 arXiv.org 在一些学科中是科研交流的中心门户，但是它对所有的学科并不是可行或可取的。

尽管我们预料 arXiv 将在它涵盖的学科领域中应用得更加广泛，但我们相信这一发展肯定是发生在有计划和有策略的方式中。arXiv 的一个原则是任何学科的扩展必须得到学术团体的支持，满足 arXiv 的质量标准，并认真考虑其业务能力和财务要求。

## 6.5　了解 arXiv 和自动下载论文的方法

请通过发送一封电子邮件信息到 arxiv-support-updates-L-request@cornell.edu 加入 arXiv 支持声明邮件列表。主题行保持空白。正文只写一个：join。

# 7 个人捐款

## 7.1 捐款方式

您可以通过康奈尔大学校友和朋友网站向 arXiv 捐款。您所捐款项 100% 将用来资助有助于 arXiv 全球科研团体的发展和新项目。

## 7.2 支持项目

请参照 arXiv 路线图的特殊项目部分。像我们在最近的一份年度更新中所描述的那样，目前的经营模式在 190 个会员组织、西蒙斯基金会和康奈尔大学图书馆的支持下，在覆盖维护成本方面进展得非常顺利。然而，为了资助超越常规工作的新项目我们需要增加额外的资源。

## 7.3 关于捐款的其他问题

（1） arXiv 的历史和影响如何？

请参考这篇新闻报道，其强调了 arXiv 的重要性：arXiv

获得 1000 万的提交量 [arXiv Hits 1 Million Submissions（2015-01-12），见 https://www.library.cornell.edu/about/news/press-releases/arxiv-hits-1-million-submissions-0]。

（2）arXiv 的可持续发展计划是什么？

关于我们目前的预算、管理模式、优先权和年度报告信息，请参考我们的经营模式和可持续发展计划。

（3）arXiv 目前的运营预算如何？

请参考我们的预算。

（4）我认为我一直是资助 arXiv 的，但是为什么会收到康奈尔大学的付款页？

arXiv 是由康奈尔大学图书馆管理的。处理我们的特殊项目的工作流程是由康奈尔校友事务与发展管理的。您的贡献将支持且仅支持 arXiv。

（5）我正在进行在线捐赠，arXiv 会用我的个人信息做些什么？

您的地址和电话号码将被独立地用来确认您的信用卡信息，不会添加到联系人列表中。无论是 arXiv.org 还是康奈尔大学图书馆都不会保留这些信息。

（6）我正在进行在线捐赠，为什么我会被问"In Honor/Memory"信息？

在线捐赠门户是由康奈尔校友事务和发展管理的，arXiv.org 中一些领域是不需要捐赠的，"In Honor/Memory"就在其中。如果您喜欢请随意捐赠"In Honor/Memory",或者通过点击"继续"按钮(在右下方)跳过这一步。

(7)我所在的机构目前支持 arXiv 吗?

详见我们的捐赠者。

(8)我所在的机构目前支持 arXiv,为什么我也要支持?

我们的成员机构很慷慨地支持我们目前的运营预算。近 5 年的经营计划表现了一个基线维护方案。它的发展基于 arXiv 2010—2012 年基本支出分析。它不考虑任何新的功能性要求或其他不可预见的资源需求。虽然建立了发展储备基金用来资助这些开支,但它不足以通过盈余资金资助重大发展努力。

(9)我怎样能够获得 arXiv 的最新消息?

请发送邮件到 arxiv-support-updates-L-request@cornell.edu:

● 主题栏保持空白。

● 正文:join。

(10)我的捐赠可以免税(美国)吗?

可以。

(11)康奈尔的税务号码(美国)是?

150532082。

(12)我可以拿到收据吗?

可以。捐赠后您将收到一个感谢的页面。感谢页面就是收据。

(13)我能从美国以外的国家进行捐赠吗？

可以。康奈尔接受国际信用卡，arXiv 欢迎来自世界各地的支持！

(14)我还有其他的问题，想要与 arXiv 团队的成员进行讨论。

您可以通过邮件联系我们：support@cornell.edu。我们欢迎您的问题和意见。

# 8 arXiv 策略回顾

## 8.1 概要

从用户的视角来看，arXiv 一直是一个成功的、杰出的学科知识库系统，为全世界众多科学家的需求服务。然而，这一服务背后面临着重大的压力。科学咨询委员会和会员咨询委员会 2015 年年度会议的结论显示，除了目前的经营模式重点关注系统维护，arXiv 团队需要为筹款付出重大的努力，寻求赠款和合作。我们需要首先创建一个有吸引力的、连贯的愿景，使我们能有说服力地表达我们基础运营的基金筹集目标超出了目前的可持续发展计划。我们想利用即将到来的 arXiv 25 周年作为参与我们一系列愿景设置活动的一个重要的里程碑。

## 8.2 愿景设置和资金筹集的策略与时间轴

(1) 2015 年 11—12 月

确定一系列考虑要做愿景设置的区域，并创建一个测量措施（已完成）。

进行科学咨询委员会/会员咨询委员会调查，民意调查他们对未来方向和优先事项的意见（已完成）。

启动用户调查计划进程（已完成）。

聘用一名兼职人员（或将此角色分配给现有的员工）与我们合作，协调规划活动（已完成）。

(2) 2016年1—2月

分析并汇报愿景设置实践的研究结果（已完成）。

启动以增加年度贡献为目标的2016年度筹款活动（arXiv的会员咨询委员会隶属小组将提出建议，以增加年度会员收入，通过白金会员模式）（已完成）。

计划并试验一项基于首次愿景设置实践的用户调查，考虑到不同的领域：作为读者的用户、作为提交者的用户、作为管理员的用户（已完成）。

收集来自会员咨询委员会IT小组关于建立一个知识库专家讨论组来考虑arXiv当前的基础设施和工作流的目标和预期结果的意见，并且推行一个举办一场研讨会的计划（计划会议Ⅰ）（已完成）。

(3) 2016年3—5月

开展用户调查（已完成）。

主持IT外部的专家/顾问来对我们的未来方向提出建议（计划会议Ⅰ）（已完成附加信息）。

启动愿景设置进程、用户研究和IT讨论组结果分析和汇报（已完成）。

(4) 2016 年 6—7 月

准备一份报告，总结用户调查的主要发现（已完成）。

基于愿景设置进程、用户研究和 IT 讨论组的结果，开始确定工作重点和资金来源来支持下一代 arXiv 的发展努力。

重复网上筹款活动（"打赏"按钮），以产生额外的资金。

基于用户研究和 IT 讨论组的结论，考虑举办一次小型会议（10～12 人），讨论一下我们的初步研究结果，并考虑未来和 arXiv 科学交流的不断变化的性质（计划会议 II：arXiv@25）（决定不举行这样的会议，因为团队已经聚集了重大的投入）。

面向 arXiv 的支持者们，设计并实施一项调查（新的战略补充，2016 年 6 月 10 日）。

(5) 2016 年 8—9 月

继续确定资金来源，初步接触。

计划并主持科学咨询委员会/会员咨询委员会年度会议，审查调查结果和新的工作计划和发展战略，推动我们前进。

(6) 2016 年 10—12 月

拨款愿景的写作与创造的工作计划（我们的希望是，在过去几个月的规划活动中，我们已经有资金来源的想法，甚至可能已经做了一些初步的工作）。

# 9 arXiv用户调查报告

## 9.1 执行摘要

作为25周年愿景设置进程的一部分，康奈尔大学的arXiv团队于2016年4月进行了一份用户调查，向全球用户社群寻求当前的服务和未来的发展方向。我们很高兴收到36 000份回复，代表着arXiv多样的社群现状（详见附录A）。当前信息表示，用户对当前所提供的服务感到满意。95%的受访者表示他们对arXiv非常满意或满意；72%的受访者表示arXiv应该关注它的主要目标——使科学论文快速获取，这将在未来充分地保持arXiv的价值。这一主题在开放文本注释上得到了普遍的反映。有很大一部分受访者建议坚持核心使命，使arXiv的合作伙伴和相关服务的提供商继续建立新的服务和创新。

很多评论反映了用户对arXiv很满意和感谢。一些用户提到了服务对他们个人职业发展的意义，并对arXiv继续存在表示感激。例如，一个典型的评论是："感谢这些年来很多人的努力，我的职业生涯没有你们的努力将会非常不同。"arXiv 也收到了很多对通过开放获取系统推进了研究传播的喝彩。一位用户提到，服务就像"科学传播的灯塔"。一些评论者表达了 arXiv 对他们个人的影响是多么关键，使他们能够快速检索到他们所在领域的最新的研究进展。整体感知，arXiv 在传统出版转型的发展方面是一个重要的领导者。没有大型机构的独立研究人员和可能会延迟获取文件的人，尤其强调了 arXiv 对他们工作的重要性。

多项选择的结合（见附录 B）和全面、深思熟虑的开放文本评论指出了一些需要升级和增强的领域。用户表示现存有限的搜索能力使他们遇到大量的阻碍，尤其是在作者搜索方面，因此提高搜索功能成为了优先要解决的问题。为提交和链接研究数据、代码、演示文稿和其他论文相关材料提供更好的支持成为另一个需要扩展的重要服务。无论他们的主题领域是什么，用户对继续实施质量控制方法，如文本查重、对提交文稿进行正确分类、拒绝没有太大科学价值的文档和要求作者解决格式相关问题等方面达成了共识。一些用户提出了将公告和邮件顺序随机化的需要。还有若干改进识别系统的需要，以及希望提供更多审核过程和政策信息的评论。

关于 arXiv 在科学出版中的角色，一些用户鼓励 arXiv 团队通过添加功能，如大胆考虑同行评议和鼓励覆盖期刊等方法并进一步推进开放获取（和新型的出版模式）。另外，许多用户强调了坚持主要使命并且不被传统出版岔开的重要性。有一个相似的意见分歧，即有关通过添加评级和注释功能鼓励开放评议过程。当为 arXiv 添加新功能来促进开放时，普遍的看法是，任何功能都需要非常认真地、系统地实施，并不危及 arXiv 的核心价值。

虽然许多受访者花时间对未来的功能增强或当前服务的实施提出建议，但一些用户反对任何改变。所有的建议，无论什么主题，都一致呼吁警惕任何变化，不要把 arXiv 变成"社会媒体"似的平台。arXiv 作为它的存在是好的，但也存在一些需要改进的地方，太多的变化可能会削弱效果和 arXiv 的总体任务。

## 9.2 主要的调查结果

### 9.2.1 提高当前 arXiv 的服务

当问及改进特定范围服务的重要性时，超过 70% 的受访者说改进搜索功能来获得更精确的结果对于不同的使用年数、年龄组、发表文章数、乡村地区和学科领域都是非常重要 / 重要的。许多评论者要求增强功能，如作者搜索、日期限制搜索、非英语语言搜索等。搜索同样是有问题的，无论用户是否搜索了一个已知的文件，浏览一个主题类别，或寻找特定的作者。

有一系列有关提升提交过程的问题，尤其是：(1) 支持研究数据、代码、幻灯片和其他材料的提交；(2) 提高研究数据、代码、幻灯片等与文章的链接；(3) 更新 TeX 引擎和其他各种各样的增强功能。每一项都有约 40% 的受访者认为非常重要/重要。对开放文本的回应同样显示出：对附加材料的支持高度重视，但不代表受访者是否认可 arXiv 或其他组织的有关服务。很多受访者支持 arXiv 整合或链接到其他服务（尤其是 GitHub），而相当数量的受访者也对长期获取及链接到 arXiv 以外的内容表示怀疑。一些受访者表达了对 arXiv 所需改进资源的关心。还有一些人表示了对将图表的数据涵盖进 arXiv 文章的兴趣。

受访者关于服务和改进的建议还有：

论文发表版本的信息和链接的一致性。

设置提醒时更精确，无论是 email 还是 RSS。一些受访者明确要求设置特定作者的邮件提醒，还有一些人对含链接的 HTML 格式的 email 表示感兴趣。

更新并保持现有的 arXiv 的 TeX 引擎，提供 TeX 模版或格式文件使提交更加简便。

通过引文和书目中可操作的链接将文献链接到一起。

允许提交 PDF，提高文件大小限制（通常有链接到图片的特殊要求），允许同时下载多个文件。

允许从作者的平台（如 Overleaf 或 Authorea）直接提交。

提供使用数据，如文章下载和阅读量。

更大比例的近期 arXiv 用户（5 年或更少）针对目前的服务更新问题选择了"没有意见"。对于这一部分问题，都有同样的趋势：一大部分近期用户表示他们没有意见，这一比例随着使用年度的增加而减少。有趣的是，这一趋势在年龄组方面没有发现，即我们的数据没有显示有高比例的年轻用户没有意见。

### 9.2.2 质量控制措施的重要性

arXiv 的用户被问及一系列关于质量控制措施的问题。基于 26 430 位受访者的回答，最重要的有（根据非常重要 / 重要排名）：

| | |
|---|---|
| 检查文章重复率，即抄袭 | 77% |
| 确认提交的文档已被正确分类 | 64% |
| 拒绝没有科学价值的文献 | 60% |
| 拒绝自我剽窃现象 | 58% |

有很大一部分人认为检查抄袭是重要的，一小部分人认为检查自我剽窃是重要的。其他的措施在统计结果中没有明显的差异。同样地，自我剽窃现象也被提到需要改进。一些人指出，上下文是关键，例如，会议论文是一个常见的典型领域。

一些受访者说，他们没有明确知道什么叫质量控制措施已经完全到位，而且感觉这一过程太不透明。其他人承认回绝没有价值的论文——"怪人写的怪文章"——与回绝其他可能不常见的匿名理由的文章之间平衡的困难。然而，即使是面对这样的批评，arXiv 当前的质量控制过程还是收到了强烈的用户满意度，用户劝告不要在另一个方向走得太远。

一些用户更倾向于接受一个更加开放的同行评议和（或）自我调节的过程，而其他人则坚持认为当前的控制给了 arXiv 自由和开放获取的速度，这是传统出版无法获得的。

总体而言，质量控制尽管重要，然而对于 arXiv 实际达到这些目标的用户评论则各有不同。就像一位受访者写到的，"对质量控制的评判是一个相对的问题"。

### 9.2.3 增加新的学科类别

73% 的受访者对于 arXiv 增加新的学科类别不感兴趣。26% 的受访者想要看到 arXiv 增加新的学科类别，并提议增加：化学（881 人）、工程学（483 人）、生物学（429 人）、经济学（248 人）、哲学（220 人）和社会科学（106 人）。还有一些小的类别，比如机器学习（82 人）和人工智能（27 人）。

一个经常回复的主题是 arXiv 不需要特别关注增加学科，应该重点关注细化和添加学科子领域，尤其是高能物理和数学学科。

### 9.2.4 开发新的服务

报告询问用户为 arXiv 提议新的服务。在对回复进行排名时发现，超过 63% 的用户认为在参考文献中增加论文链接

（参考提取）非常重要/重要。超过57%的用户认为引文输出的格式如BibTex、RIS非常重要/重要，并且为arXiv引文提取到BibTeX入口同样有超过55%的受访者提议。引文分析工具大体上有53%的受访者认为非常重要/重要。

在开放文本的评论中，在提升引文分析能力的需求方面产生了意见分歧。用户通常支持引文工具，这些用户中的很多人指出其他系统已经在这样做了，这也足以满足他们的需求。

在多项选择调查中，关于"提供一个评级系统以供用户推荐他们认为有价值的论文"的回复中，认为非常重要/重要的和认为不重要/不应该这样做的比例相当，均为36%。这与支持和不确定的分歧现象相匹配。同时，有50%的近期用户和28%的经验丰富的用户认为是有价值的。此外，有很大比例的年轻用户认为这是重要的（他们中有42%的人在30岁以下），有28%的60岁以上的用户也这样认为。虽然在开放文本的评论中产生了意见分歧，但是总体上来讲受访者对于这一提议是犹豫不决的。一些用户喜好"理想化情况"的评级，但不一定认为arXiv适合它。其他人对它会淡化arXiv的使命或它在arXiv现有体制下并不可行表示了顾虑。然而，即使是用户直接支持评级系统，也出现了是否对公众开放、同行评议、匿名评级等问题。一些受访者强调这些功能实施起来要非常认真。

诸如提供一个评级系统这样的问题，关于增加一个注释功能以供读者评论文章的提议也出现了意见分歧。34.89%的受访者认为这一功能非常重要/重要，34.08%的受访者认为不重要/不应该这样做。在开放文本回复中，反对意见的趋势和一些回复反映了强烈的消极情绪。那些支持或持无所谓态度的受访者通常都添加了警告，通常即使是积极的回应也有谨慎的感觉。一个普遍关注的主题是必须具备一套合适的系统和可核实的账户，以避免"什么话都能说（free-for-all）"。与提供评级系统问题不同的是，在不同人口特征的基础上，没有明显的差异。

### 9.2.5 搜索 arXiv 论文

arXiv 绝大多数用户访问论文直接在主页访问（79%），其次是使用 Google 搜索（50%），还有 Google 学术（35%）。

一旦将文章放置在主页上，由于使用和导航的简易性，用户行为就会混淆。32%的受访者对它的评级是简单，只有25%的受访者认为较容易找到它，21.6%的用户认为使用起来稍微困难。

为了发现内容，63%的用户会转到新的或最近的特定类别下的链接，同样有63%的用户使用 arXiv 的搜索引擎并键入一个具体的 arXiv ID、作者名或搜索词。一小部分用户（14%）依赖于每日邮件列表，然后在搜索字段中查找一篇特定的文章。

在开放文本讨论中，关于用户界面的意见有了分歧。大部分的受访者不喜欢过时的风格，但是有一小部分人肯定现存界面的简洁性，这些用户的感受有助于arXiv有效执行其使命。主要问题除了主页的外观，还有链接的数量、布局和查找提交信息。理解arXiv导航的挑战是结构组织中缺乏层次。

有关UX增强的请求包括论文的读者个性化需求，例如，"喜爱"一篇文章的能力，负责一个私人图书馆，当用户访问网站时看到评论。其他的用户提到了发展APIs来进一步促进期刊覆盖。一些用户同样建议建立一个可移动友好型版本。

很多评论者都描述了他们如何依靠其他服务与arXiv互动（arXiv定点搜索、ADS、INSPIRE）或根据他们与其他信息系统的经验推荐新的功能。他们经常称赞的有ADS、INSPIRE、谷歌学术、gitxiv.com和arxiv-sanity.com。

有关arXiv：arXiv是一个开放获取科学数字保存系统，是由西蒙斯基金会、康奈尔大学图书馆和来自全世界的大约190所成员图书馆共同资助的。该网站是由研究团体和机构共同管理和支持的，他们同时直接受益于arXiv，确保了透明化的和可持续发展的来源。它是一个有组织的学术交流论坛，由科学家和它所服务的科学文化通知和引导。截至2016年6月，arXiv已包含超过1 110 000份电子文档。2015年，

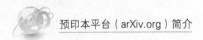

这一机构知识库就收到了 105 000 份文档提交,并且全世界就有近 13.9 亿下载量。

## 附录 A：调查对象统计

我通过以下方式使用 arXiv:（请选择所有适用的）

| 回答 | 比例 | 数量 |
| --- | --- | --- |
| 我是 arXiv 的读者 | 93% | 31 862 |
| 我是 arXiv 的作者 | 53% | 18 270 |
| 我是 arXiv 的提交者 | 50% | 17 189 |
| 我是 arXiv 的其他类型用户，请描述 | 2% | 845 |

我出版/提交给 arXiv 的文章数量是：

| 回答 | 比例 | 数量 |
| --- | --- | --- |
| 1 篇 | 11.99% | 2570 |
| 2 篇 | 8.96% | 1920 |
| 3～4 篇 | 15.19% | 3254 |
| 5～10 篇 | 23.06% | 4941 |
| 10 篇及以上 | 40.80% | 8743 |
| 合计 | 100% | 21 428 |

我目前的职业是：（请选择所有适用的）

| 回答 | 比例 | 数量 |
| --- | --- | --- |
| 我在一所学院或大学担任学术工作者（教授） | 27% | 8868 |
| 我在一所学院或大学担任学术工作者（研究人员或博士后） | 22% | 7207 |
| 我是一名非营利或政府机构的研究人员 | 8% | 2707 |
| 我是一名硕士生/博士生 | 30% | 9890 |
| 我是一名大学生 | 5% | 1514 |
| 我是（请描述） | 13% | 4353 |

4353名受访者中，有13%的人显示出了不同的职业类别。比例最高的包括公司或工业界的研究人员（900人）、工程师（515人）和退休的个人（478人）。还有一些受访者描述自己是科学写作者、编辑或自由编辑。其他的回复类型包括数据科学家、自定义业余研究者、自定义外行、失业、教师和一般好奇者（如一个人将研究作为爱好）。

作为一名用户,我主要的兴趣学科领域是:(请选择所有适用的)

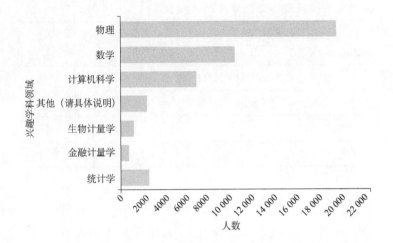

大约有 2000 名受访者选择了"其他"选项来指定他们的主要兴趣区域。人数最多的是天体物理学(726 人)和天文学(653 人)。

我使用 arXiv 已经有:

| 回答 | 比例 | 数量 |
| --- | --- | --- |
| 0~2 年 | 19.54% | 6470 |
| 3~5 年 | 28.96% | 9592 |
| 6~10 年 | 25.44% | 8425 |
| 11 年及以上 | 26.06% | 8632 |
| 合计 | 100% | 33 119 |

我的年龄:

| 回答 | 比例 | 数量 |
|---|---|---|
| 30 岁以下 | 37.42% | 12 364 |
| 30～39 岁 | 31.27% | 10 332 |
| 40～49 岁 | 13.76% | 4545 |
| 50～59 岁 | 9.30% | 3073 |
| 60～69 岁 | 5.77% | 1908 |
| 70 岁及以上 | 2.47% | 817 |
| 合计 | 100% | 33 039 |

我主要的工作地点位于:

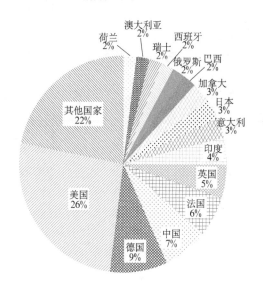

其他国家:113 个国家中占 1% 或者更少比例。

## 附录B：对arXiv目前服务与未来方向的观点

提高arXiv以下的服务有多重要？

| 问题 | 非常重要&重要 | 稍微重要 | 不重要&不应该这样做 | 无所谓 |
|---|---|---|---|---|
| 提高检索功能来获得精确的结果（例如，通过附加检索词来缩小检索结果范围，通过出版年或者所属机构筛选等） | 70.38% | 19.34% | 6.14% | 4.13% |
| 为提交研究数据、代码、幻灯片和其他与论文相关的材料提供支持（例如，我想要上传我的论文数据集/机器可视化表格） | 41.95% | 22.64% | 14.03% | 21.37% |
| 为链接研究数据、代码、幻灯片和其他与论文相关的材料提供支持（例如，我想要在SlideShare上能够链接我的幻灯片） | 40.65% | 25.20% | 17.70% | 16.45% |

续表

| 问题 | 非常重要&重要 | 稍微重要 | 不重要&不应该这样做 | 无所谓 |
|---|---|---|---|---|
| 为通过更新 TeX 引擎来提交研究论文提供支持 | 39.36% | 23.17% | 16.71% | 20.76% |
| 提升邮件提醒系统以供读者可以自定义他们的设置，并选择接收特定的子主题的通知 | 37.85% | 26.48% | 20.25% | 15.42% |
| 提升引用机制（链接文章到博客和评论引用论文） | 36.52% | 29.50% | 20.30% | 13.67% |
| 通过提供清晰的指示和简单的语言来简化提交过程 | 32.45% | 22.55% | 25.20% | 19.80% |

开发以下 arXiv 的服务有多重要？

| 问题 | 非常重要&重要 | 稍微重要 | 不重要&不应该这样做 | 无所谓 |
|---|---|---|---|---|
| 在参考文献中加入直接的文章链接（支持参考文献提取） | 63.04% | 26.89% | 5.78% | 4.29% |
| 提供格式化的引文输出如 BibTeX、RIS 等 | 57.68% | 23% | 10.95% | 8.37% |
| 在 arXiv 引文中提取 BibTeX 输入 | 55.54% | 23.82% | 9.72% | 10.91% |

续表

| 问题 | 非常重要 & 重要 | 稍微重要 | 不重要 & 不应该这样做 | 无所谓 |
|---|---|---|---|---|
| 提供引文分析工具（研究论文被引频次和模式的研究） | 52.95% | 27.08% | 14.28% | 5.69% |
| 支持遵守公共/开放获取授权（资助需要将研究数据公开的机构政策）通过允许论文的最终版本提交时涵盖资助来源和编号信息 | 42.06% | 26.21% | 13.68% | 18.05% |
| 允许上传到 arXiv 的同时提交文章到期刊 | 39.28% | 23.09% | 25.23% | 12.40% |
| 提供一个评级系统供读者推荐他们认为有价值的文章 | 36.28% | 21.76% | 35.56% | 6.40% |
| 允许链接（互操作）到其他知识库（如图书馆运营），这样一篇文章被 arXiv 接收的同时也可以被其他知识库接收 | 35.25% | 28.14% | 17.25% | 19.36% |
| 开发一个注释功能，让读者评论论文 | 34.89% | 23.62% | 34.08% | 7.41% |

你去哪里搜索 arXiv 的文章？（请选择所有适用的选项）

| 回答 | 比例 | 数量 |
|---|---|---|
| 直接去 arXiv.org (arXiv 主页) | 79% | 22 804 |
| ADS | 14% | 4144 |
| Inspire | 13% | 3773 |

## 9 arXiv 用户调查报告

续表

| 回答 | 比例 | 数量 |
| --- | --- | --- |
| 谷歌学术 | 35% | 10 016 |
| 谷歌搜索引擎 | 50% | 14 440 |
| arXiv 邮件提醒 | 14% | 4086 |
| 其他搜索引擎 | 5% | 1402 |
| arXiv 的学科门户，如 Math Front | 4% | 1203 |
| 其他（请详细描述） | 9% | 2662 |

如果你使用 arXiv 主页来搜索文章，它的导航性能怎么样？

| 回答 | 比例 | 数量 |
| --- | --- | --- |
| 非常简单 | 14.85% | 3916 |
| 简单 | 32.05% | 8450 |
| 稍微简单 | 25.20% | 6644 |
| 稍微困难 | 21.60% | 5696 |
| 困难 | 5.02% | 1324 |
| 非常困难 | 1.27% | 336 |
| 合计 | 100% | 26 366 |

如果你使用过 arXiv 主页，你通常怎样导航我们的页面？（请选择所有适用的选项）

| 回答 | 比例 | 数量 |
|---|---|---|
| 在特定类型下点击"new"或者"recent" | 63% | 16 503 |
| 使用 arXiv 搜索引擎并键入一个具体的 arXiv-id、作者名称或搜索词 | 63% | 16 478 |
| 收取每日邮件提醒，然后在搜索领域找到一篇具体的文章 | 14% | 3692 |
| 其他，请描述 | 3% | 853 |

以下当前的质量控制措施的重要性如何？

| 问题 | 非常重要 & 重要 | 稍微重要 | 不重要 & 不应该这样做 | 无所谓 |
|---|---|---|---|---|
| arXiv 进行文本查重：作者使用了大量与其他作者的文章相同的文本，且没有说明这些不是他们自己的材料，即"抄袭" | 77.41% | 14.66% | 4.96% | 2.96% |
| arXiv 确保提交的文章已正确分类（学科分类在 arXiv 主页上） | 64.38% | 25.32% | 7.01% | 3.29% |
| arXiv 拒绝没有科研价值的文章 | 60.02% | 19.14% | 15.49% | 5.35% |

续表

| 问题 | 非常重要 & 重要 | 稍微重要 | 不重要 & 不应该这样做 | 无所谓 |
|---|---|---|---|---|
| arXiv 检查作者使用了过多自己早期研究内容,即"自我剽窃"(重复使用个人的相同内容且没有标记引用) | 57.77% | 24.64% | 14.08% | 3.51% |
| arXiv 检查相关格式问题(文章的行数、参考文献遗漏、过大的文本等)并在文章公布之前告知作者进行修改 | 55.00% | 29.83% | 11.51% | 3.66% |
| arXiv 在允许添加链接之前管理科学内容的引用(链接到博客和评论) | 39.60% | 26.31% | 17.59% | 16.50% |

总体来说,您对 arXiv 的满意度如何?

| 回答 | 比例 | 数量 |
|---|---|---|
| 非常满意 | 52.92% | 14 770 |
| 满意 | 42.43% | 11 841 |
| 稍微满意 | 3.55% | 990 |
| 稍微不满意 | 0.54% | 150 |
| 非常不满意 | 0.15% | 42 |
| 无所谓 | 0.42% | 116 |
| 合计 | 100% | 27 909 |

预印本平台(arXiv.org)简介

以下哪一项最恰当地描述了您对arXiv需要怎样向前发展的意见?

| 回答 | 比例 | 数量 |
|---|---|---|
| arXiv应该重点关注它的主要目标,即将科学论文尽快可获取。这些已足够让arXiv在未来保持它的价值 | 71.94% | 19 865 |
| arXiv应该扩大它的主要使命,将更多的时间和资源用来提供新的服务。这些对于arXiv在未来保持它的价值是十分必要的 | 19.59% | 5410 |
| 无所谓 | 8.47% | 2340 |
| 合计 | 100% | 27 615 |